THE SOLDIER'S POCKET COMPANION

1746

The Naval & Military Press Ltd

published in association with

ROYAL
ARMOURIES

Published by
The Naval & Military Press Ltd
Unit 10 Ridgewood Industrial Park,
Uckfield, East Sussex,
TN22 5QE England
Tel: +44 (0) 1825 749494
Fax: +44 (0) 1825 765701
www.naval-military-press.com

in association with

In reprinting in facsimile from the original, any imperfections are inevitably reproduced and the quality may fall short of modern type and cartographic standards.

THE
Soldier's Pocket-Companion,

OR THE

Manual Exercise of our British Foot,

As now practisd by his Majesty's *special Command;*

With previous Directions to

Officers, in Regard to their *proper Salutes* to the KING, *or any of the* Royal Family, &c

To which is Added

A

Short View of the Use of the

SMALL-SWORD.

MDCCXLVI.

Sold by the Proprietor *Bloe*, Engraver & Copper-Plate Printer, *the Corner of Kings Head Court Holborn.*

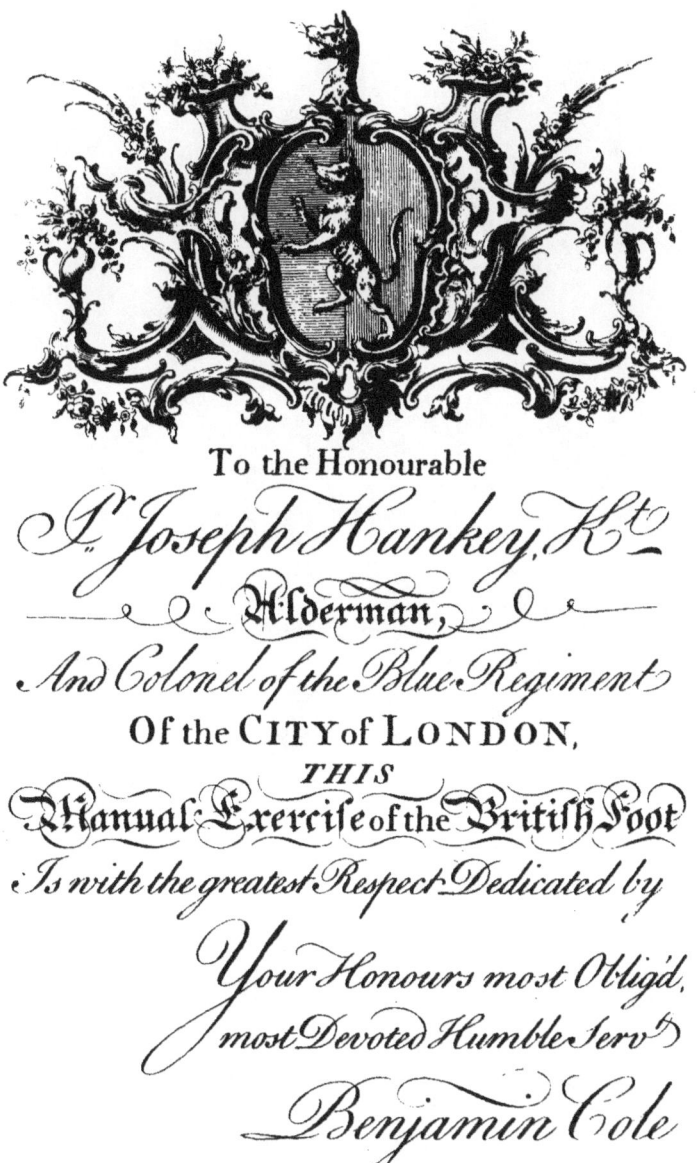

To the Honourable

S.ʳ Joseph Hankey, K.ᵗ

Alderman,

And Colonel of the Blue Regiment
Of the CITY of LONDON,
THIS
Manual Exercise of the British Foot
Is with the greatest Respect Dedicated by

Your Honours most Oblig'd,
most Devoted Humble Serv.ᵗ

Benjamin Cole

The TABLE.

Officers Salutes.

	Words of Command.	No. of Motions.	Pages.
	The Standing Salute	6	1 to 7
	The Marching Salute	7	8 to 15

Manual Exercise.

	Take Care		15
1	Join your Right Hand to your Firelock	1	17
2	Poise your Firelock	1	18
3	Join your Left Hand to your Firelock	2	19 and 20
4	Cock your Firelock	2	21
5	Present	1	
6	Fire	1	22
7	Recover your Arms	1	
8	Half Cock your Firelock	2	
9	Handle your Primer	3	23 to 25
10	Prime	2	26
11	Shut your Pan	4	
12	Cast about to Charge	2	27 and 28
13	Handle your Cartridge	3	29
14	Open your Cartridge	2	30
15	Charge with Cartridge	3	
16	Draw your Rammer	4	31 to 34
17	Shorten your Rammer	3	35 to 37
18	Put it in the Barrel	6	38 to 43
19	Ram down your Charge	2	
20	Recover your Rammer	3	43
21	Shorten your Rammer	3	
22	Return your Rammer	6	
23	Cast off your Firelock	3	44
24	Your Right Hand under the Lock	1	
25	Poise your Firelock	1	45
26	Shoulder your Firelock	3	46
27	Rest your Firelock	4	
28	Order your Firelock	3	47 to 49
29	Ground your Firelock	4	50 to 53
30	Take up your Firelock	4	53
31	Rest your Firelock	3	54 to 55
32	Club your Firelock	4	56 to 59
33	Rest your Firelock	4	59

34 Secure your Firelock	3	60 and 61
35 Shoulder your Firelock	5	} 61
36 Poise your Firelock	2	
37 Rest on your Arms	3	62 to 64
38 Draw your Bayonet	2	65 and 66
39 Fix your Bayonet	5	67 to 69
40 Rest your Bayonet	3	70
41 Charge your Bayonet Breast high	4	71 to 73
42 Push your Bayonet	2	74
43 Recover your Arms	2	75
44 Rest your Bayonet on your Left Arm	2	76 and 77
45 Rest your Bayonet	3	
46 Shoulder your Firelock	4	} 78
47 Present your Arms	4	
48 To the Right (4 Times)	3	
49 To the Right about	3	} 79
50 To the Left as you were	3	
51 To the Left (4 Times)	3	
52 To the Left about	3	
53 To the Right as you were	3	} 84
54 Poise your Firelock	1	
55 Rest on your Arms	3	
56 Unfix your Bayonet	3	85
57 Return your Bayonet	4	86 and 87
58 Poise your Firelock	3	} 87
59 Shoulder your Firelock	4	

Total Number of Motions 189

Small Sword.

1 The Guard	88
2 The Thrust in Carte	89
3 The Thrust in Tierce	90
4 The Carte Thrust under the Arm	91
5 The Thrust in Seconde	92
6 Tierce Thrust on Carte Side	93
7 Carte Thrust over the Arm	94
8 The Flanconnade	95
9 The Pass in Tierce	96

I

The Officer muſt ſtand upright and bold, with his Half Pike in his Right Hand, about the ſame Diſtance from his Right Foot as he can extend his Right Arm, holding his Half Pike full in his Right Hand in a direct Line from his Right Shoulder, as far as his Arm will permit without Conſtraint, taking Care that the Half Pike be upright and his Left Hand on his Left Side juſt above the Hip, Thumb behind, Fingers before.

The Officer's Salute at ye Head of his Troop.

ATTITUDE. I.

ATTITUDE. II.

The Standing Salute.

1st Motion.

Fall back with your Right Foot and Hand, and at the same Time seize your Half Pike with your Left Hand, about two Foot and a half from the Ferrel, keeping both your Arms extended from your Body as far as you can without Constraint, and your Aspect as much as possible towards the Front.

The Standing Salute.

2d Motion.

Quit your Right Hand, and at the same Time lifting up your Half Pike with your Left Hand, seize it again with your Right close to the Ferrel, your Thumb and Fingers extended, your Elbow also a little bent and extended; bring up your Right Foot at the same Time to the Hollow of the Left or to a *Roman* T) but not too close but so as to stand firm.

ATTITUDE. III.

IV

ATTITUDE. IV.

The Standing Salute.

3d Motion.

Fall back again, with your Right Foot, and lifting up your Elbow, let the Spear of your Half Pike drop within two or three Inches of the Ground; the Staff falling over the Back of the four Fingers of your Left Hand, which muſt be about the Heighth of your Breaſt, with both Arms and Fingers extended.

The Standing Salute.

4th Motion.

Bring up your Right Foot again, near the Hollow of the Left, and at the same Time bring your Half Pike to a Recover.

ATTITUDE. V.

VI

The Officer's Salute upon a March.
ATTITUDE. I.

The Standing Salute.

5th Motion.

Fall back again with your Right Foot, and at the same Time quit your Right Hand, and seize your Half Pike as high as you can towards the Spear, that when you come to Order again it may be in a Right Line from your Right Shoulder.

The Standing Salute.

6th Motion.

Quit your Left Hand, and at the same Time bring up your Right Hand Pike, and Right Foot together; placing the But End of your Half Pike, and Right Foot, at once on the Ground together, to your proper Front; in a direct Line with your Left Foot, and your Right Arm in a Line from your Right Shoulder, keeping your Pike perpendicular; then seize your Hat briskly with your Left Hand, and bring it down with a brisk Motion by your Side, as low as your Arm will extend.

VII

ATTITUDE II.

VIII

ATTITUDE. III.

The Officer muſt March, till he comes within about 20 Paces of the Perſon he is to Salute; with his Half Pike comported as in the Figure.

The Marching Salute.

1ſt Motion.

With a briſk Motion fling off your Right Hand to the Right, with your Half Pike, and turning the Ferrel of your Half Pike foremoſt, bring it on your Right Shoulder, your Right Elbow ſquare, your Left Hand ſet againſt your Left Side, juſt above the Hip, Thumb behind, Fingers before; and the Spear drooping behind, a little lower than the But End.

ATTITUDE. IV.

X

ATTITUDE V.

The Marching Salute.

2d Motion.

Upon ſtepping forward with your Right Foot, caſt off your Half Pike, in a direct Line from your Shoulder, as far as poſſible without Conſtraint; then ſtepping forward with the Left Foot, at the ſame Time, ſeize your Half Pike, with your Left Hand, within two Foot and a half of the Ferrel.

The Marching Salute.

3d Motion.

Bring up the Right Foot again, opposite to the Hollow of the Left; and at the same Time, seize your Half Pike, with your Right Hand, at the Ferrel, your Half Pike perpendicularly upright before you, or in a proper Recover.

ATTITUDE VI.

XII

ATTITUDE. VII.

The Marching Salute.

4th Motion.

Stepping forward with the Left Foot, lift up your Elbow, and let the Spear of your Half Pike drop within two or three Inches of the Ground; the Staff falling over the Back of the four Fingers of the Left Hand, your Body upright, both Arms and Fingers extended.

The Marching Salute.

5th Motion.

Stepping forward with your Right Foot, bring it up near the Hollow of the Left; or, to a Roman T: And at the same Time, bring up your Half Pike, perpendicularly before you, or to a Recover.

ATTITUDE. VIII.

XIV

ATTITUDE IX.

The Marching Salute.

6th Motion.

On the next Step with the Left Foot, quit the Ferrel of the Half Pike, with your Right Hand, and seize it again with it; about the Middle of the Staff: Fling it off with a straight Arm.

The Marching Salute.

7th Motion.

Stepping forward with the Right Foot, bring it to your Right Shoulder: keeping your Right Elbow square Take off your Hat with your Left Hand, and bring it to your Left Side, by a quick Motion.

XV

The Ensign upon a March.

XVI

The Manual Exercise, &c.
TAKE CARE.

Take Care.

As soon as the Word of Command is given, you must observe a profound Silence, and make no Motion either with your Head, Body, Feet or Hands, but such as shall be ordered, looking to the Officer who is to give the Word of Command, carrying your Firelock straight on your Shoulder, Barrel up, Muzzle high, pressing the Guard to your Breast, your Feet a Step Distance, the Heels in a Line, and your Toes turned out.

N. B. *This shews likewise the last Motion of Shoulder your Firelock, as in the* 26*th,* 35*th,* 46*th, and* 59*th Words of Command.*

1. *Join your Right Hand to your Firelock.*

Your Firelock being carried in the forementioned Posture upon the Left Shoulder, you must turn it inwards with the Left Hand, the But to be sunk a little, and at once take hold with the Right Hand behind the Lock, both Elbows in an equal Line, but not constrained.

N. B. *This Figure shews the 1st Motion of the 27th, the 33d, and 36th Words of Command; and the Barrel being suppos'd upwards, (as in the last Figure) instead of the Lock: It likewise shews the 2d of the 26th, the 4th of the 35th, the 3d of the 46th, and the 2d of the 59th Words of Command.*

XVII

Join your Right Hand to your Firelock.

Poise your Firelock.

2. *Poise your Firelock.*

At the Word of Command, with both Hands and a quick Motion bring up the Firelock from your Shoulder, at the same Time thrust it from you with your Right Hand; in doing which, let your Left Hand fall down by your Side, the side Plate opposite to your Roller, with your Arm a little bended, the Lock turned outwards, and the Thumb inwards, against the Face, and your Feet in the same Posture as when shoulder'd.

N. B. *This Figure likewise shews the 2d Motion of the 29th, the 1st of the 34th, the 2d of the 35th, the 2d of the 36th, the 3d of the 46th, and the 3d of the 58th Words of Command.*

3. *Join your Left Hand to your Firelock.*

1st Motion.

Turn your Firelock the Barrel towards you, at the same Time seize it with the Left Hand, so that the Little Finger touch the Lock; holding your Firelock in both Hands, with your Arms extended as much as possible without Constraint; tell 1, 2.

N. B. *This Figure likewise shews the* 2d *Motion of the* 27th *and* 33d *Words of Command.*

XIX

Join your Left Hand to your
Firelock.

Rest your Firelock.

3. *Join your Left Hand to your Firelock.*

2d Motion.

With a quick Motion bring your Firelock down, the But opposite to the Right Knee, the Muzzle pointing a little forwards, the Stock in the Left Hand, with your Right Thumb on the Cock, the Forefinger before the Trigger, and the other Fingers behind the Guard. At the same Time that you bring down your Firelock, you must step a little back with your Right Foot, the Toe pointing to the Right; the Right Knee stiff, the Left Knee a little bending, and your Body very straight, and face to the Front as much as possible.

N. B. *This is the Rest, when fac'd to the Left;* Fig. 80, *is the Front Rest;* Fig. 82, *when fac'd to the Right or Left about; and* Fig. 84, *when fac'd to the Right; which Figures shew the 4th Motion of the 27th, the 3d of the 31st, the 4th of the 33d, the 3d of the 40th, the 3d of the 45th, the 4th of the 47th, and the 3d of the 49th, 50th, 51st, 52d and 53d Words of Command.*

4. *Cock your Firelock.*

1st Motion.

Keep your Thumb upon the Cock, and bring up your Firelock with both Hands before you, the Cock, Roller high; at the same Time bring up your Right Foot, the Heel within half a Foot of the Hollow of the Left Foot, and the Toe pointing to the Right, the Firelock close to your Breast, that you may the easier bend the Cock; tell 1, 2.

2d Motion.

Cock, and at the same Time thrust your Firelock from you with both Hands, holding your Thumb upon the Cock, your Fore-Finger before the Trigger, keeping your Arms stretch'd out before your Body.

N. B. *This is the Recover when fac'd to the Left;* Fig. 79, *is the Front Recover;* Fig. 81, *when fac'd to the Right or Left about; and* Fig. 83, *when fac'd to the Right; which Figures shew the 2d Motion of the 12th, the 4th of the 21st, the 2d of the 40th, the 1st of the 41st, the 2d of the 43d, the 2d of the 45th, and the 2d and 3d of the 48th, 49th, 50th, 51st, 52d and 53d Words of Command.*

Cock your Firelock & Recover.

XXII

Present

5. *Present.*

In presenting, take away your Thumb from the Cock, and move the Right Foot a little back, the Toe turned to the Right, the Body to the Front, and place the But in the Hollow, between the Right Breast and the Shoulder, keeping the Fore-Finger before the Trigger, but without touching it, and the other three Fingers behind the Guard, the Elbows in an equal Line, the Head straight upwards, the Body upright, but a little press'd forwards against the Firelock, the Left Knee a little bent, and the Right Knee stiff.

6. *Fire.*

As soon as this Word is given, draw the Trigger briskly with the Fore-Finger, and take care to draw the Trigger but once.

7. *Recover your Arms.*

Bring up your Firelock straight before your Cock, Roller high; the Right Heel near the Hollow of your Left Foot, keeping the Posture as in Explanation, and *Fig.* 21.

8. *Half Cock your Firelock.*

1st Motion.
Bring the Firelock close to your Breast, and half bend the Cock; tell 1, 2.

2d Motion.
Thrust it from you with both Hands, as *Fig.* 21.

9. *Handle your Primer.*

1st Motion.

Fall back briskly with your Right Foot behind the Left, that the Heels come straight behind one another, the Left Toe pointing to the Front; and bring down your Firelock to the Right at the same Time with both Hands, and a quick Motion, keeping the Muzzle on a Level with the rest of the Barrel; tell 1, 2.

N. B. *This Figure shews likewise the 3d Motion of Shutting the Pan.*

Handle your Primers.
1ˢᵗ Motion.

XXIIII

Handle your Primers.
II^d. Motion.

9. *Handle your Primer.*

2d Motion.

Quitting the Firelock with the Right Hand clap your Pouch, and take hold of your Primer, the Thumb on the Spring Cover; tell 1, 2.

N. B. *This Figure likewise shews the 2d Motion of Prime.*

9. *Handle your Primer.*

3d Motion.

Bring it within two Fingers Breadth of the Pan, the Thumb upwards.

Handle your Primers.
III.^d Motion.

Prime

1.ˢᵗ Motion.

10. *Prime.*

1st Motion.
Hold your Firelock still, and turning up that Hand with the Primer, shake out as much Powder in the Pan as is necessary; let fall your Primer, and open your Hand; tell 1, 2.

2d Motion.
Throw it back behind the But End, the Palm outwards, and remain in that Posture till the following Word of Command.

11. *Shut your Pan.*

1st Motion.
Take hold of the Steel with your Thumb upwards, and your two Fore-Fingers under; tell 1, 2.

2d Motion.
Shut your Pan; tell 1, 2.

3d Motion.
Seize your Firelock with your Right Hand behind the Lock; (as in *Fig.* 23.) tell 1, 2.

4th Motion.
Bring up your Firelock to the Recover, as *Figure* 21.

N. B. *If in this Figure the Fingers are supposed over the Steel of the Pan, it will shew the two first Motions.*

12. *Caſt about to Charge.*

1ſt Motion.

Turn the Firelock with both Hands, the Barrel outwards; tell 1, 2.

XXVII

Cast about to Charge.
1ˢᵗ Motion.

Cast about to Charge.
II^d Motion.

12. *Caſt about to Charge.*

2d Motion.

Let go the Right Hand, bringing down the Firelock with the Left; ſtep forwards with the Right Foot, tho' not directly before the Left; but place it a little to the Right, that the Body may preſent itſelf the better forwards; taking hold of the Muzzle with the Right Hand, that the bringing down of the Firelock, the moving of the Right Foot and the taking hold of the Muzzle, be done at the ſame Time; hold it with your Right Hand, the Thumb upwards near the Rammer, and the Barrel downwards, keeping the Body ſtraight, only the Right Knee a little bent, which muſt remain ſo till you have charged.

N. B. *This Figure likewiſe ſhews the 1ſt Motion of the 13th, and the 2d and 3d of the 23d Words of Command.*

13. *Handle your Cartridge.*

1st Motion.

Bring the Firelock with both Hands to your Body; tell 1, 2.

2d Motion.

Quit your Firelock with your Right Hand, holding it with your Left Hand in a Ballance, the Muzzle pointing a little forward, and at the same Time clap your Pouch, and take hold of your Cartridge; tell 1, 2.

3d Motion.

Bring it within one Inch of the side of the Muzzle, the Thumb upwards, and the Right Elbow square.

XXIX

Handle your Cartridge.
II.ᵈ Motion.

Open your Cartridge.
1ˢᵗ Motion.

14. *Open your Cartridge.*

1st Motion.

Bring the Cartridge to your Mouth, and bite off the Top, finking your Elbow; tell 1, 2.

2d Motion.

Bring it again to its former Place, holding it with the Thumb upwards.

15. *Charge with Cartridge.*

1st Motion.

Bring the Cartridge juft before the Muzzle, turning up your Hand and Elbow, and fix it in at the fame Time; tell 1, 2.

2d Motion.

Raife your two Fore-Fingers; tell 1, 2.

3d Motion.

Clap them on the Muzzle brifkly, and remain fo with the Elbow fquare.

16. *Draw your Rammer.*

1st Motion.

Seize the Rammer with your Fore-Finger and Thumb of your Right Hand, the Thumb upwards; tell 1, 2.

Draw your Rammer.
1ˢᵗ Motion.

Draw your Rammer.
II.ᵈ Motion.

16. *Draw your Rammer.*

2d Motion.

Draw it out as far as your Arm will reach; tell 1, 2.

N. B. *This Figure likewise shews the 1st Motion of Recover your Rammer.*

16. *Draw your Rammer.*

3d Motion.

Take hold of it close to the Stock, turning the Thumb downwards; tell 1, 2.

N. B. *This Figure shews likewise the 2d Motion of Recovering the Rammer.*

XXXIII

Draw your Rammer.
III.ᵈ Motion.

Draw your Rammer.
IV.th Motion.

16. *Draw your Rammer.*

4th Motion.

Draw it quite out, holding it between the Thumb and the two Fore-Fingers, the whole Arm ſtretch'd out in a Line with the Right Shoulder; the ſmall End towards you, and the other from you in an even Line.

N. B. *This Figure ſhews likewiſe the 3d Motion of Recovering the Rammer.*

17. *Shorten your Rammer.*

1st Motion.

Move the middle Finger, which supports the Rammer, and turn it quick with the thick End down, and hold it so in your Hand, with an out-stretch'd Arm, in a Line with your Shoulder, the Thumb upwards; tell 1, 2.

N. B. *This Figure likewise shews the 1st Motion of the 21st Word of Command, only the thick End of the Rammer is upwards.*

XXXV

Shorten your Rammer.
1st Motion.

Shorten your Rammer.
II.ª Motion.

17. *Shorten your Rammer.*

2d Motion.

Set the thick End againſt the lower Part of your Breaſt; tell 1, 2.

N. B. *This Figure ſhews likewiſe the 2d Motion of the 21ſt Word of Command.*

17. *Shorten your Rammer.*

3d Motion.

Slip your Hand down to a Hand's Breadth of the End, the Rammer in a Line with the Barrel, the Thumb upwards, and the Elbow a little turn'd out from the Body.

N. B. *This Figure likewise shews the 3d Motion of the 21st Word of Command.*

Shorten your Rammer
III.d Motion.

Put it in the Barrell.
1.ᶠᵗ Motion.

18. *Put it in the Barrel.*

1st Motion.

Bring the Rammer a little above the Muzzle, and place the thick End on the Cartridge; then tell 1, 2.

N. B. *This Figure likewise shews the 1st Motion of Returning the Rammer, only there the small End is put into the Stock instead of the thick End into the Barrel.*

18. *Put it in the Barrel.*

2d Motion.

Thrust it down as far as your Hand will permit; tell 1, 2.

N. B. *This Figure likewise shews the 2d Motion of Returning the Rammer, only with the Difference noted in the 1st Motion.*

Put it in the Barrell.
II.^d Motion.

Put it in the Barrell.
III^d Motion.

18. *Put it in the Barrel.*

3d Motion.

Seize it about the Middle; then tell 1, 2.

N. B. *This Figure likewise shews the 3d Motion of Returning the Rammer, with the Difference noted in the 1st Motion.*

18. *Put it in the Barrel.*

4th Motion.

Thrust it down as before; tell 1, 2.

N. B. *This Figure likewise shews the 4th Motion of Returning the Rammer, with the Difference noted in the 1st Motion.*

Put it in the Barrell
IV.th Motion.

Put it in the Barrell.
v.th Motion.

18. *Put it in the Barrel.*

5th Motion.

Seize it at the Top; tell 1, 2.

N. B. *This Figure likewise shews the 5th Motion of Returning the Rammer, only in that the Palm of the Hand is put at the Top of the Rammer.*

18. *Put it in the Barrel.*

6th Motion.

Thruſt it down to your Hand, holding the Rammer faſt with the Thumb upwards.

19. *Ram down your Charge.*

1ſt Motion.

Draw the Rammer as far as the Arm unforc'd will permit; tell 1, 2.

N. B. *Fig.* 42. *ſhews this Motion.*

2d Motion.

Ram down the Charge with an ordinary Force, hold the Rammer as before.

N. B. *This Fig. ſhews this Motion.*

20. *Recover your Rammer.*

1ſt Motion.

Draw your Rammer with a quick Motion till Half of it be out of the Barrel; (as *Fig.* 32.) tell 1, 2.

2d Motion.

Seize it cloſe to the Muzzle with the Thumb downwards; (as *Fig.* 33.) tell 1, 2.

3d Motion.

Draw it quite out of the Barrel, holding it with the Thick End towards your Shoulder, obſerving the ſame Poſition, as in Explanation and *Fig.* 34.

21. *Shorten your Rammer.*

1ſt Motion.

Turn down the ſmall End of the Rammer with your two Fore Fingers and Thumb; (as *Fig.* 35.) tell 1, 2.

2d Motion.

Set it againſt your Breaſt; (as *Fig.* 36.) tell 1, 2.

3d Motion.

Slip your Hand within a Foot of the End; as *Fig.* 37.

Put it in the Barrell.
VIth Motion.

Cast off your Firelock
1.ˢᵗ **Motion.**

22. *Return your Rammer.*

1st Motion.
Bring the small End with a gentle Turn under the Barrel, and place it in the Stock; (as *Fig.* and *Note* 38.) tell 1, 2.

2d Motion.
Thrust it in as far as your Hand will permit; (as *Fig.* 39.) tell 1, 2.

3d Motion.
Seize it in the Middle; (as *Fig.* 40.) tell 1, 2.

4th Motion.
Thrust it down as before; as *Fig.* 41. tell 1, 2.

5th Motion.
Set the Palm of your Hand against the thick End; (as *Fig.* and *Note* 42.) tell 1, 2.

6th Motion.
Thrust it quite down.

23. *Cast off your Firelock.*

1st Motion.
Extend your Right Arm to the Right, in a Line with your Shoulder; tell 1, 2.

2d Motion.
Take hold of your Firelock, your Thumb even with the Muzzle: as *Fig.* 28. tell 1, 2.

3d Motion.
Thrust your Firelock from your Body; as Explanation and *Fig.* 28.

24. *Your Right Hand under the Lock.*

Face on the Left Heel to the Left, at the same Time turning the Muzzle directly up, you seize the Firelock with the Right Hand behind the Lock, holding the Firelock from your Body, and your Hands as low as you can without Constraint.

25. *Poise your Firelock.*

Face very quick on the Left Heel to the Right, and at the same Time bring the Firelock with the Right Hand before you, letting your Left Hand fall down by your Side, pushing the Firelock suddenly with the Right Hand forwards, the Arm a little bended, so that the thrusting forwards of the Firelock, and the setting down of the Right Foot, be done at the same Time; as *Fig.* 18.

XLV

Your Right Hand under the Lock.

XLVI

Shoulder your Firelock.
1ᶠᵗ Motion.

26. *Shoulder your Firelock.*

1st Motion.

Turn your Firelock with the Right Hand, the Barrel outwards, and the Guard inwards, againſt the Left Shoulder; at the ſame Time ſeize the But with your Left Hand, placing your Thumb in the Hollow; tell 1, 2.

2d Motion.

Bring it with both Hands upon the Left Shoulder without moving your Head, and keep both Elbows in a Line; (as *Fig.* and *Note* 17.) tell 1, 2.

3d Motion.

Quit your Right Hand, letting it fall down by your Side, ſinking your Left Elbow at the ſame Time; (as *Fig.* 16.)

27. *Reſt your Firelock.*

1st Motion.

Join your Right Hand; as in Explanation and *Fig.* 17.

2d Motion.

Come to your Poiſe; as in Explanation and *Fig.* 18.

3d Motion.

Seize your Firelock with your Left Hand; as in Explanation and *Fig.* 19.

4th Motion.

Come down to your Reſt; as in Explanat. and *Fig.* 20.

28. *Order your Firelock.*

1st Motion.

Slip up your Left Hand as high as your Right Shoulder; bring back at the same Time your Right Hand towards your Right Thigh, holding your Firelock perpendicular; tell 1, 2.

XLVII

Order your Firelock.
1st Motion.

Order your Firelock.
II.d Motion.

28. *Order your Firelock.*

2d Motion.

Let go the Right Hand, sinking the Firelock with the Left; at the same Time seize your Firelock with the Right Hand near the Muzzle, that the Thumb be upwards and even with it; tell 1, 2.

28. *Order your Firelock.*

3d Motion.

Quit your Left Hand, and fit down the But End of the Firelock upon the Ground even with your Toe, at the Outfide of your Right Foot, and perform it with that Quickneſs, that your Right Foot and the Firelock come down at the ſame Time, the Heels in a ſtraight Line, the Toes turned outwards, letting your Right Arm hang from the Hand to the Elbow by the Side of the Firelock, and the Left Hand hanging by the Left Side.

N. B. *This Figure likewiſe ſhews the 4th Motion of Taking up the Firelock.*

XLIX

Order your Firelock.
III.d Motion.

Ground your Firelock
1st Motion.

29. *Ground your Firelock.*

1ſt Motion.

Lift up your Right Foot, and making a half Face to the Right, place it againſt the flat End of the But, and at the ſame Time turn the Barrel of your Firelock towards your Body; tell 1, 2.

29. *Ground your Firelock.*

2d Motion.

Step directly forward with the Left Foot, flipping your Right Hand to the Middle of the Barrel, your Left Hand hanging down, and at the same Time you bring down your Right Knee on the Firelock, looking up; tell 1, 2.

N. B. *This Figure shews likewise the 2d Motion of Taking up the Firelock.*

Ground your Firelock.
II.ᵈ **Motion.**

Ground your Firelock.
III.^d Motion.

29. *Ground your Firelock.*

3d Motion.

Raiſe your Self again, ſtepping back with your Left Foot, and keeping your Body half fac'd to the Right; tell 1, 2.

N. B. *This Figure ſhews likewiſe the 1ſt Motion of Taking up the Firelock.*

29. *Ground your Firelock.*
4th Motion.
Turn your Right Foot on the Heel over the But End, and bring in your Body to its proper Front, letting both Arms hang down by your Sides.

30. *Take up your Firelock.*
1st Motion.
Turn your Right Foot on your Heel over the But End of the Firelock, and set it down behind the same, making a half Face to the Right; extend your Right Arm a little to your Right Side; (as *Fig.* 52.) tell 1, 2.

2d Motion.
Step forward with the Left Foot along the Firelock; at the same Time take hold of it by the Middle of the Barrel with an out-stretch'd Arm and a stiff Body; (as *Fig.* 51.) tell 1, 2.

3d Motion.
Raise up yourself and the Firelock again; bringing back the Left Foot; then tell 1, 2.

4th Motion.
Lift up your Right Foot again and set it at the Inside of the But, slipping up your Right Hand as high as the Muzzle, and turning the Barrel towards the Right Shoulder; stand in the Posture that is shewn in Explanation and *Fig.* 49.

N. B. *You must observe in Grounding your Firelock not to keep your Hand on the Muzzle, but to sink them to the Middle of the Barrel; and in taking it up, to take hold at the same Place, then also slip your Hand up to the Muzzle with Ease.*

Note, *It is further to be observ'd, that at the Grounding, and Taking up your Firelock, you must keep up your Head.*

Ground your firelock.
IV.th Motion.

LIV

Rest your Firelock.
I.ˢᵗ Motion.

31. *Reſt your Firelock.*

1ſt Motion.

Turn your Thumb inwards, and ſlip your Hand as low as the Arm will permit without Conſtraint; tell 1, 2.

31. *Rest your Firelock,*

2d Motion.

Raise your Firelock with the Right Hand, taking hold of it at the same Time with the Left, just under the Right; tell 1, 2.

3d Motion.

Let go your Right Hand, and place it behind the Lock, stepping back with your Right Foot at the same Time, so that the resting your Firelock, and stepping back with the Right Foot, be done at once; then keep your Firelock, Body, and Feet in the same Posture, as in Explanation and *Fig.* 20.

Rest your Firelock.
II.ᵈ Motion.

LVI

Club your Firelock.
1ˢᵗ Motion.

32. *Club your Firelock.*

1st Motion.

Keep your Firelock firm in your Left Hand, and cast it about with your Right; bring up the Right Foot at the same Time, and take hold of it with the Right Hand as low as you can without Constraint, the Guard right against your Eyes, the Muzzle and Left Thumb downwards, and the Lock from you; tell 1, 2.

N. B. *This Figure likewise shews the 3d Motion of the Return from the Club to the Rest, only then the Barrel is from you.*

32. *Club your Firelock.*

2d Motion.

Let go the Left Hand, and place it at the End of the Stock, raising the Firelock at the same Time with the Right Hand, and keeping it with out-stretch'd Arms opposite to the Left Shoulder; tell 1, 2.

N. B. *This Figure likewise shews the 2d Motion of returning from the Club to the Rest, only the Barrel from you instead of the Lock.*

Club your Firelock.
II.d Motion.

LVIII

Club your Firelock.
III.ᵈ Motion.

32. *Club your Firelock.*

3d Motion.

Bring it on the Left Shoulder with the Lock upwards; tell 1, 2.

N. B. *This Figure likewise shews the 1st Motion of the Return from the Club to the Rest; only there the Barrel is upwards instead of the Lock.*

32. *Club your Firelock.*

4th Motion.

Quit your Right Hand with a quick Motion, and let it hang down by your Right Side.

33. *Rest your Firelock.*

1st Motion.

Turn your Firelock with the Left Hand inwards sinking your Firelock, and at the same Time take hold with your Right Hand a Handful above the Left, the Elbows in an Equal Line; (see *Fig.* and *Note* 58.) tell 1, 2.

2d Motion.

Bring it with both Hands before your Body, the But high, and your Arms extended; (see *Fig.* and *Note* 57.) tell 1, 2.

3d Motion.

Let go your Left Hand, and sink your Firelock with the Right, let the Guard be even with your Eyes, seizing it at the same Time, near the Lock, with your Left Hand turned, the Thumb downwards; (see *Fig.* and *Note* 56.) tell 1, 2.

4th Motion.

Let go your Right Hand, and turn the Firelock with the Left, bringing the But End down, and come to your Rest, stepping back with your Right Foot; as Explanation and *Fig.* 20.

Club your Firelock.
IV.th Motion.

Secure your Firelock.
II.^d Motion.

34. *Secure your Firelock.*

1ſt Motion.

Come briſkly to your Poiſe, (as *Fig.* 18) tell 1, 2.

2d Motion.

Seize the Firelock with your Left Hand a Handful from the Lock, turning the Barrel outwards, and bringing the Firelock oppoſite to your Left Shoulder, the Muzzle directly up; tell 1, 2.

N. B. *The 1ſt Motion in returning from the Secure to the Shoulder, differs from this, in being brought cloſe to the Body, and the Lock outwards.*

34. *Secure your Firelock.*
3d Motion.

Quit your Right Hand, and bring the Firelock with the Left Hand under your Left Arm, the Lock betwixt the Wrist and the Elbow, the Barrel downwards, and the Muzzle a Foot from the Ground.

35. *Shoulder your Firelock.*
1st Motion.

Bring the Firelock with a quick Motion before you, the Muzzle upwards, and the Lock turn'd outwards, and seize it at the same Time with the Right Hand under the Lock; (see *Fig.* and *Note* 60.) tell 1, 2.

2d Motion.

Thrust it from you with the Right Hand, and let go the Left, at the same Time come to your Poise; (as *Fig.* 18.) tell 1, 2.

3d Motion.

Clap your Left Hand to the But, the Thumb in the Hollow; (as Expla. and *Fig.* 46.) tell 1, 2.

4th Motion.

Lay it on your Shoulder; as Expla. 46. tell 1, 2.

5th Motion.

Quit your Right Hand; as in Explan. 46. and *Fig.* 16.

36. *Poise your Firelock.*

This is done as in Explanat. and *Fig.* 17, 18.

LXI

Secure your Firelock.
III.ᵈ Motion.

Rest on your Arms.
1ˢᵗ Motion.

37. Rest on your Arms.

1st Motion.

Sink your Firelock as low as you can without Constraint with your Right Hand, seizing it at the same Time with your Left, the Height of your Chin, the Left Elbow turn'd out; tell 1, 2.

37. *Reſt on your Arms.*

2d Motion.

Seize the Muzzle with your Right Hand, tell 1, 2.

Rest on your Arms.
II.ᵈ Motion.

Rest your Arms
III.ᵈ Motion.

37. *Reft on your Arms.*

3d Motion.

Bring the But to the Ground, flipping up your Left Hand at the same Time, close to your Right.

38. *Draw your Bayonet.*

1st Motion.

Seize your Bayonet with the Right Hand, the Thumb in Hollow; tell 1, 2.

Draw your Bayonet.
1st Motion.

LXVI

Draw your Bayonet.
II^d Motion.

38. *Draw your Bayonet*.

2d Motion.

Draw it out briskly, facing full to the Right, with an extended Arm, the Point of the Bayonet straight up, with your Thumb in the Hollow of the Shank, that the Notch of the Socket may come even with the Sight of the Barrel, when you fix it on the Muzzle.

39. *Fix your Bayonet.*

1st Motion.

Turn briskly up with Foot and Hand to the proper Front, placing the Socket of the Bayonet on the Muzzle; tell 1, 2.

2d Motion.

Thrust it down as far as the Notch will permit; tell 1, 2.

LXVII

Fix your Bayonet.
1.st Motion.

LXVIII

Fix your Bayonet.
III.d Motion.

39. *Fix your Bayonet.*

3d Motion.

Turn it from you, and fix it; tell 1, 2.

39. *Fix your Bayonet.*

4th Motion.

Cast your Hand a little to the Right, with a square Elbow; tell 1, 2.

5th Motion.

Seize your Firelock with the Palm of your Right Hand on the Back of your Left, as *Fig.* 64.

Fix your Bayonet
IV.ʰ Motion.

Rest your Bayonet.
1st Motion.

40. *Rest your Bayonet.*

1st Motion.

Raise the Firelock with your Right Hand as high as your Forehead, and flip your Left Hand at the same Time as low as possible without Constraint; tell 1, 2.

2d Motion.

Raise your Firelock with your Left Hand, turning the Barrel towards you; and at the same Time seize it under the Lock, observing the Posture as Explanation, and *Fig.* 21. tell 1, 2.

3d Motion.

Then come to your Rest, as Explanation, and *Fig.* 20.

41. *Charge your Bayonet Breaſt high.*

1ſt Motion.

Bring your Firelock to the Recover; as *Fig.* 21. tell 1, 2.

2d Motion.

Throw back your Right Hand; tell 1, 2.

Charge your Bayonet Breast high.
II.d Motion.

LXXII

Charge y̓ Bayonet Breast High.
III.d Motion.

41. *Charge your Bayonet Breast high.*

3d Motion.

Clap the Palm againſt the Plate of the But, the Barrel being towards you; tell 1, 2.

41. Charge your Bayonet Breast high.

4th Motion.

Fall back with your Right Foot, your Heels in a Line; come to your Charge, having the But End in a full Right Hand, your Thumb upon it, the Barrel upwards, the Left Elbow turned out from the Body, and the Point of the Bayonet the Height of your Breast.

LXXIII

Charge ỹ Bayonet Breast High.
IV.ᵗʰ Motion.

Push your Bayonet.
1st *Motion.*

42. *Puſh your Bayonet.*

1ſt Motion.

Puſh your Bayonet forwards, without raiſing or ſinking the Point; and at the ſame Time bring the But of the Firelock before your Left Breaſt; tell 1, 2.

2d Motion.

Bring it back to its former Poſture.

43. *Recover your Arms.*

1st Motion.

Seize your Firelock with the Right Hand behind the Cock; tell 1, 2.

2d Motion.

Come up to your Recover.

Recover your Arms.
1st Motion.

LXXVI

Rest ỹ Bayonet on your left Arm.
1ˢᵗ Motion.

44. *Rest your Bayonet on your Left Arm.*

1st Motion.

Turn the Lock of the Firelock from you; tell 1, 2.

44. *Rest your Bayonet on your Left Arm.*

2d Motion.

Stepping out with the Right Foot let go your Left Hand, sink your Firelock, and at the same Time take hold of the Cock and Steel with your Left Hand, the Cock lying on your middle Finger, and the lower Joint of your Thumb on the Steel; keep both Arms as low as possible without Constraint; the But between your Thighs, and the Bayonet pointing exactly to your Left, and as far from your Shoulder, as the Situation of both your Arms and the But will permit.

LXXVII

Rest ỳ Bayonet on your left Arm.
II.ᵈ Motion.

LXXVIII.

Rest your Bayonet.
1.st Motion.

45. *Rest your Bayonet.*

1st Motion.
Slip your Left Hand without moving the Firelock, and take hold of the Stock above the Lock, your Thumb inwards; tell 1, 2.

2d Motion.
Bring the Firelock to the Recover, with your Right Heel against the Hollow of the Left Foot, as *Fig.* 21. tell 1, 2.

3d Motion.
Come briskly to the Rest.

46. *Shoulder your Firelock.*

1st Motion.
Come briskly to the Poise, as *Fig.* 18.

2d Motion.
Clap your Left Hand on the But, as *Fig.* 46.

3d Motion.
Lay the Firelock on your Shoulder, as *Fig.* 17. and Note.

4th Motion.
Quit your Right Hand, as *Fig.* 16.

47. *Present your Arms.*

This is done as in Explanation 27.

48. *To the Right,*
   To the Right,
   To the Right,
   To the Right,

1st Motion.

In each of these four Facings you first come to the Recover, as *Fig.* 21. tell 1, 2.

2d Motion.

Face upon the Left Heel to the Right, keeping your Firelock well Recover'd; tell 1. 2.

3d Motion.

Come to your Rest nimbly, stepping back with your Right Foot.

49. *To the Right about.*

This is done as in the foregoing Explanation; only you now face to the Right about.

50. *To the Left as you were.*

Observe the same Time as in the other Facings; coming briskly to the Left about.

LXXIX

The Recover in Front.

LXXX

The Rest in Front.

The Recover when Fac'd, to the Right or Left about.

LXXXII

The Rest when Fac'd to ÿ
Right or Left about.

LXXXIII

The Recover when Fac'd to the Right.

LXXXIV

The Rest when Fac'd to the Right

51. *To the Left,*
    To the Left,
    To the Left,
    To the Left,

This is perform'd like the Facings to the Right, only with this Difference, that you turn upon the Left Heel to the Left every Time, the fourth Part of a Circle; and observe, as in Explanation 48.

52. *To the Left about.*

This is done as in Explanation 50.

53. *To the Right as you were.*

This is done as in Explanation 49.

54. *Poise your Firelock.*

Come briskly to the Poise in one Motion.

55. *Rest on your Arms.*

This is done as in Explanation 37.

56. *Unfix your Bayonet.*

1st Motion.

Slip up your Bayonet with the Right Hand; tell 1, 2.

2d Motion.

Turn it towards you; tell 1, 2.

3d Motion.

Slip it quite off the Muzzle, thrusting it from you at the same Time.

LXXXV

Unfix your Bayonet.
II.^d Motion.

Return your Bayonet.
II.d Motion.

57. *Return your Bayonet.*

1st Motion.

Turn briskly to the Right on the Left Heel, with an extended Arm, and the Point of the Bayonet upwards; (as *Fig.* 66.) tell 1, 2.

2d Motion.

Sink the Point of your Bayonet, and place it in the Scabbard; tell 1, 2.

57. *Return your Bayonet.*

3d Motion.

Thrust it quite in, holding up your Head, and looking to the Right; tell 1. 2.

4th Motion.

Extend your Arm to its former Posture, and come briskly up to your proper Front, seizing the Firelock near the Muzzle, with your Right Hand above the Left.

58. *Poise your Firelock.*

1st Motion.

The same as the 1st Motion in Explanation 40. tell 1, 2.

2d Motion.

Raise the Firelock with the Left Hand, seizing with the Right Hand under the Lock; tell 1, 2.

3d Motion.

Thrust it from you, coming to the Poise.

59. *Shoulder your Firelock.*

This is done as in Explanation 26.

Return your Bayonet.
III.d Motion.

LXXXVIII

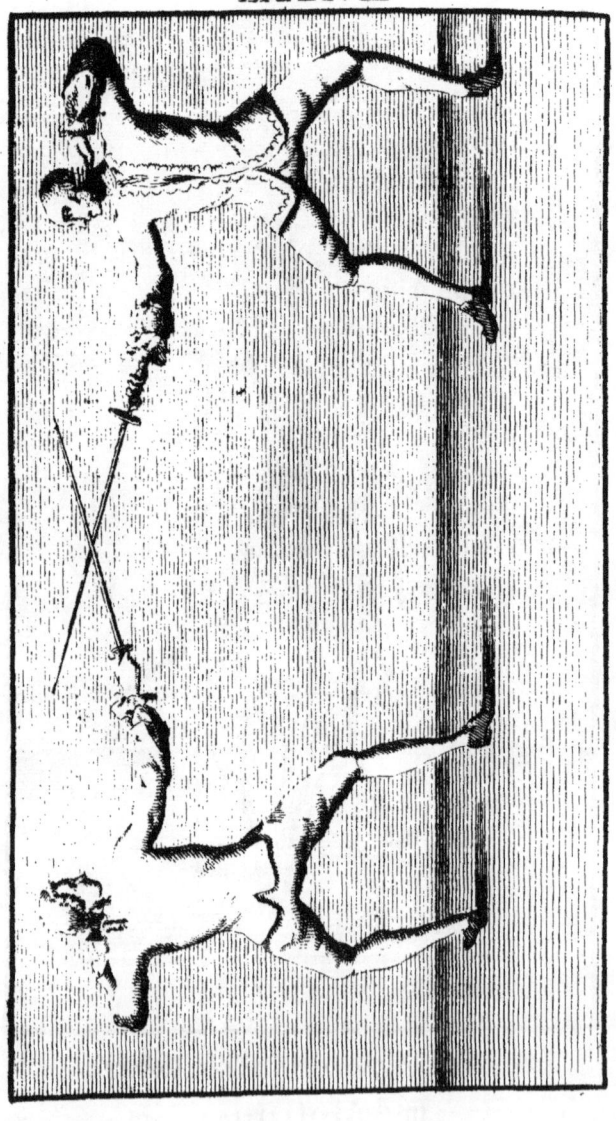

The Guard.

The Sword is to be held, steady, and firm, in the Right Hand, upon the Flat or Demy-Carte, with the Nail of the Thumb upwards: The Pummel of the Sword in a Line with the short Ribs; with the Point somewhat elevated. The Left Arm elevated, and bent in a Semi-Circle, the Elbow turn'd out; the Left Hand rais'd to the Height of the Left Eye, the Thumb downward; the Shoulders well squared, the Body upright, and well edged in a Line, with the Point of your Sword; and which should be sustain'd chiefly on the Left Foot; the Right Foot flat and firm on the Ground; the Right Leg perpendicular; and the Knee a little bent: The Left Toe turn'd outwards, so as the Heel may just be clear of the Heel of the Right Foot, being distant from each other about two Foot: The Head erect, directing your View along the Sword Arm, towards your Enemy.

2. *The Thrust in Carte.*

Having a good Guard as before, and within Measure, your Sword engag'd in Carte; you deliver the Thrust, with a full Longe; dropping the Point, the Nails of the Sword Arm turned upwards; the Wrist elevated, higher than the Shoulder, and well supported with the Arm extended: At the same Time throwing off the Left Arm, in a strait Line to the Left, with the Palm upwards; the Body almost upright, and well supported. The Right Leg perpendicular with the Knee a little bent, the Toe opposite to your Adversary, and in a Line with your Sword. The Left Leg and Thigh extended with a stiff Knee, both Feet flat and firm on the Ground; the Head erect and inclining a little over the Right Shoulder, from whence you observe your Thrust.

LXXXIX

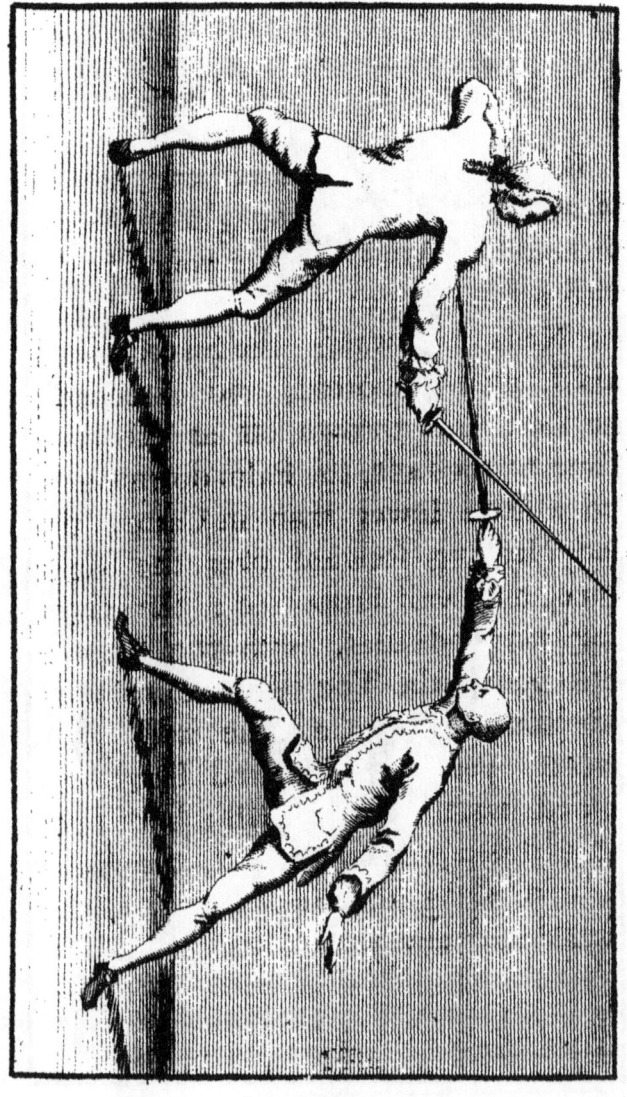

XC

3. *The Thruſt in Tierce.*

Being in a good Guard and within Meaſure, the Sword engag'd in Tierce without the Arms, deliver your Thruſt, with a full Longe; dropping the Point; the Nails downwards, the Sword Hand rais'd, well ſupported, and extended; the Left Arm thrown off at the ſame Time to the Left, the Palm downwards, and a little lower than in Carte; the Body leaning forward over the Right Knee, and well ſupported; the Right Leg perpendicular, with the Knee a little bent; the Toe oppoſite to your Adverſary, and in a Line with your Sword; your Left Leg and Thigh extended, with the Knee ſtiff; both Feet flat and firm on the Ground, the Head inclin'd along the Sword Arm, under which you view your Thruſt.

4. *Carte Thrust under the Arm.*

The Body is in the same Attitude, in delivering this Thrust, as in simple Carte; with this Difference, that it is given, under the Adversaries Sword Arm, which is held too high in Carte.

XCI

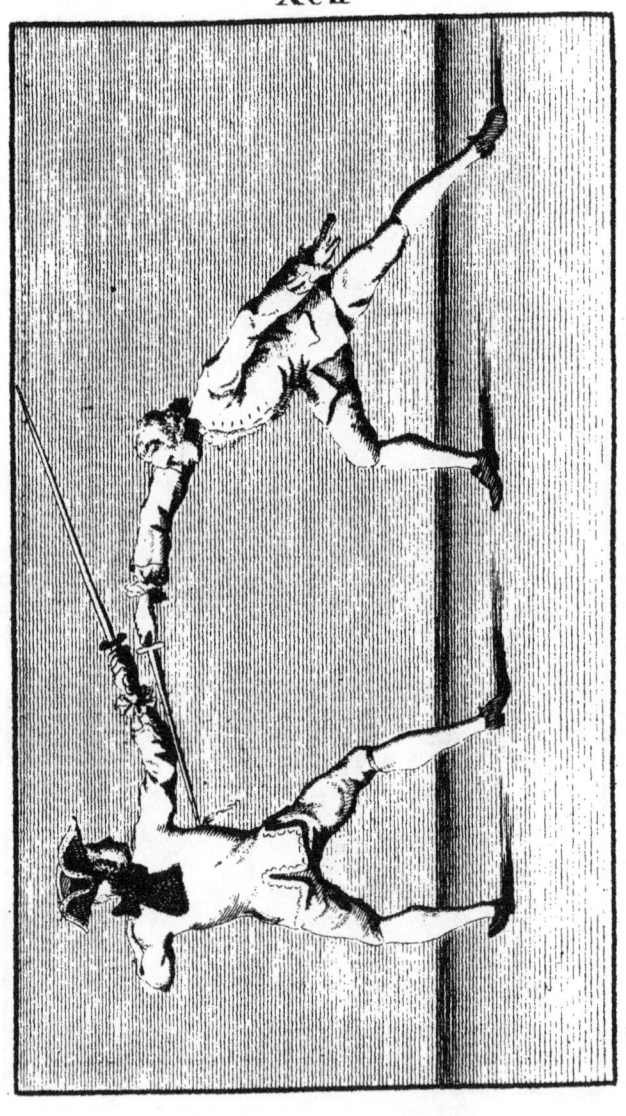

XCI

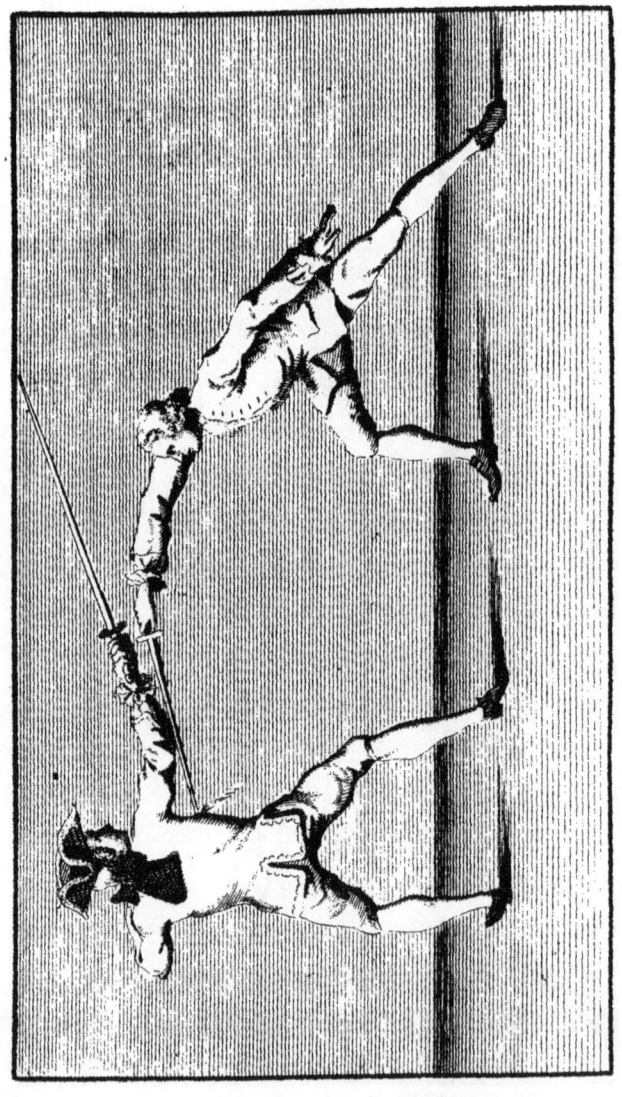

5. *The Thrust in Seconde or Tierce under the Arm.*

This Thrust is given in the same Position of Body, as in the high Tierce; only the Adversaries Sword Arm, being too high in Tierce, it is push'd under it.

6. *Tiere thruft on Carte Side.*

This Thruft is deliver'd, as in the upper Tierce, by reafon the Enemy holds his Sword in Guard, inclin'd too much to the Right.

XCIII

XCIV

7. *Carte Thrust over the Arm, or an Outside Carte.*

This Thrust is deliver'd in the same Position of the Body, as in the upper Carte within the Arms; because your Adversary, holding his Wrist too low in Tierce, inclines his Point too much to the Left.

8. *The Flanconnade.*

This Thruft is made, by gaining the Feeble of your Adverfaries Sword, (which may be done without, or upon his Longeing) fliding forward with the Point directly under your Adverfaries Wrift, towards his Flank; the Left Hand at the fame Time brought forward towards the Enemy's Sword; the Pofition of the Body as in Carte.

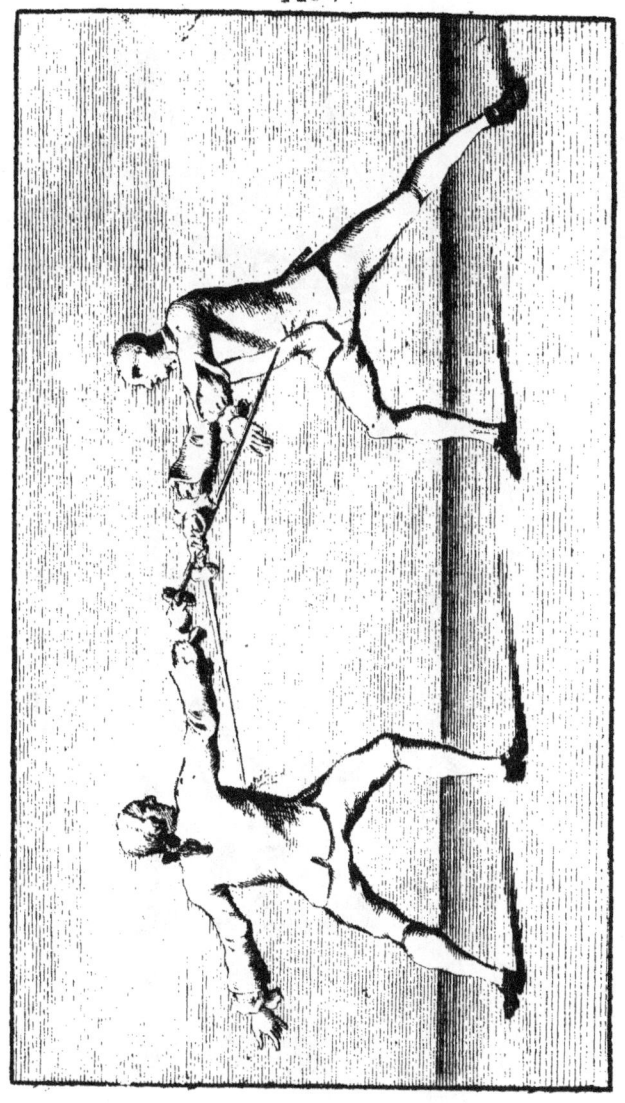

XCVI

9. *The Pass in Tierce.*

Is a Thrust deliver'd in an outside upper Tierce, with this Difference, that instead of Longeing with the Right Foot you step forward with the Left: The Left Leg is perpendicular whilst the Right Leg and Thigh are extended; the Toe of the Right Foot on the Ground, and the Heel rais'd.

N. B. *In this Thrust you double your Velocity, Force and Distance.*

www.ingramcontent.com/pod-product-compliance
Lightning Source LLC
Chambersburg PA
CBHW031144160426
43193CB00008B/245